Contents

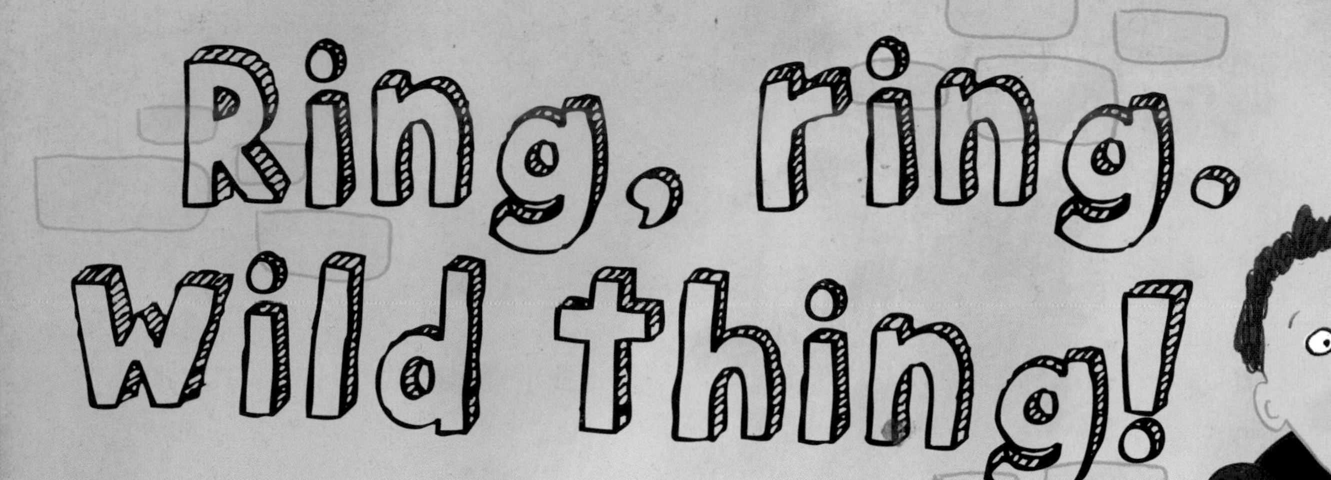

If you're WILD about animals, today's your lucky day.

There's a penguin at the door! You could invite it in...

Hmm, this could
get messy...
Cute!

Waddle!

Penguins have wings, but they can't fly.

They use their wings to help them swim.

You will need

Flippers

Face mask

An enormous swimming pool

Zoom!

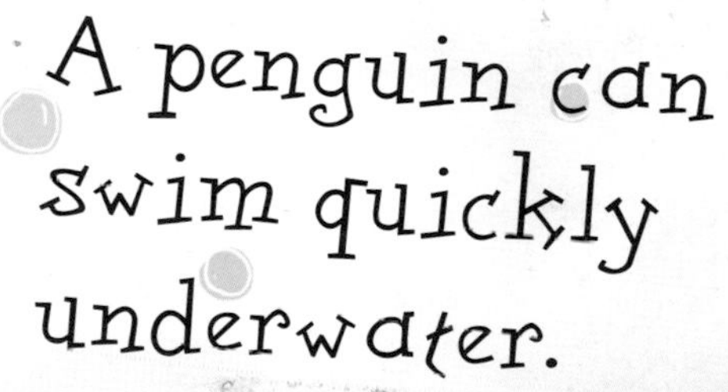

A penguin can swim quickly underwater.

Swim

Penguins spend most of their time in the ocean.

They eat and play in the water. They come to land to lay their eggs.

You will need

To move to the seaside

To love the water – as much as your pet!

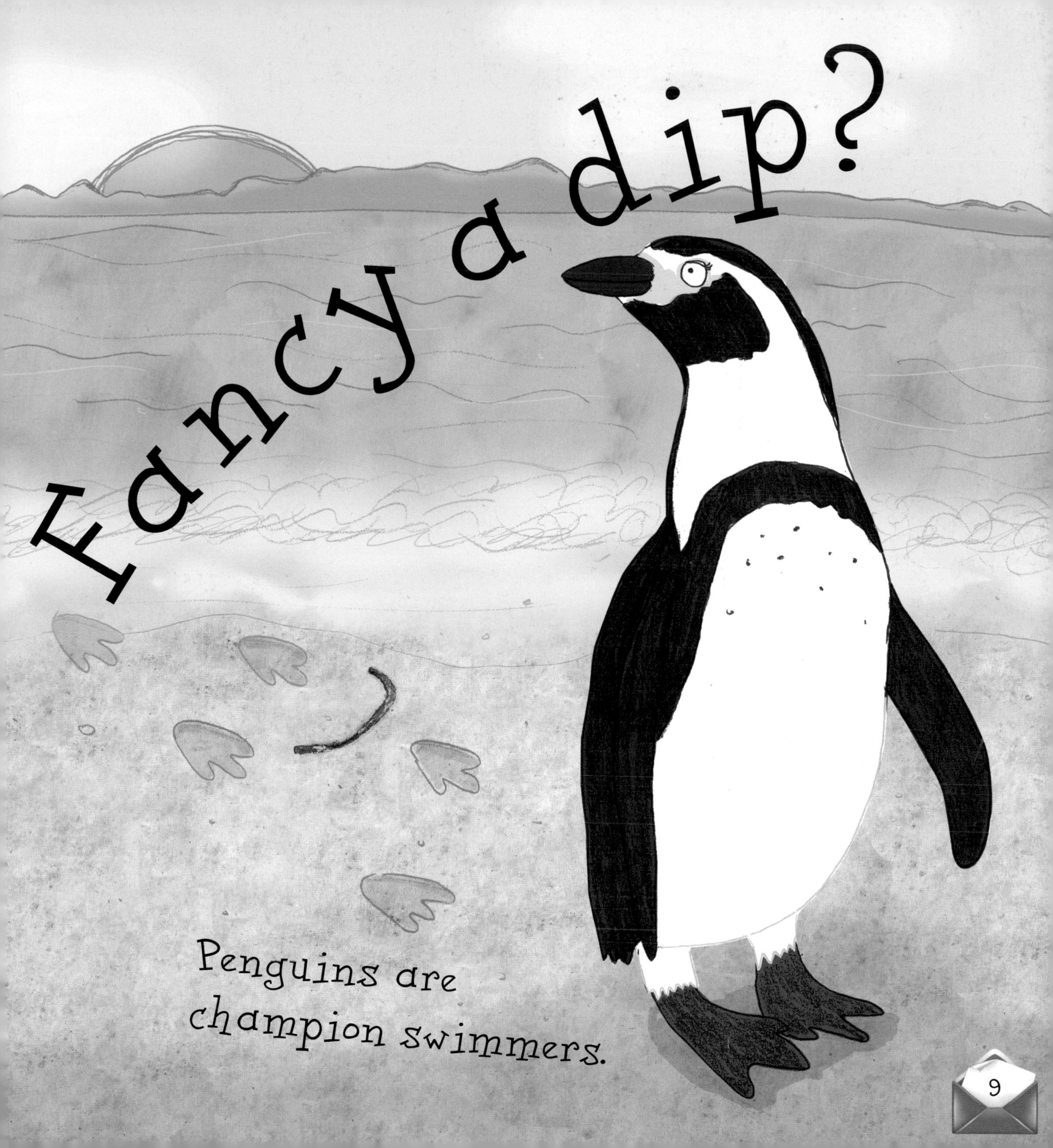

Fancy a dip?

Penguins are champion swimmers.

Smelly!

Penguins feed on fish and other sea creatures.

A strong fishy smell follows them wherever they go. Yuck!

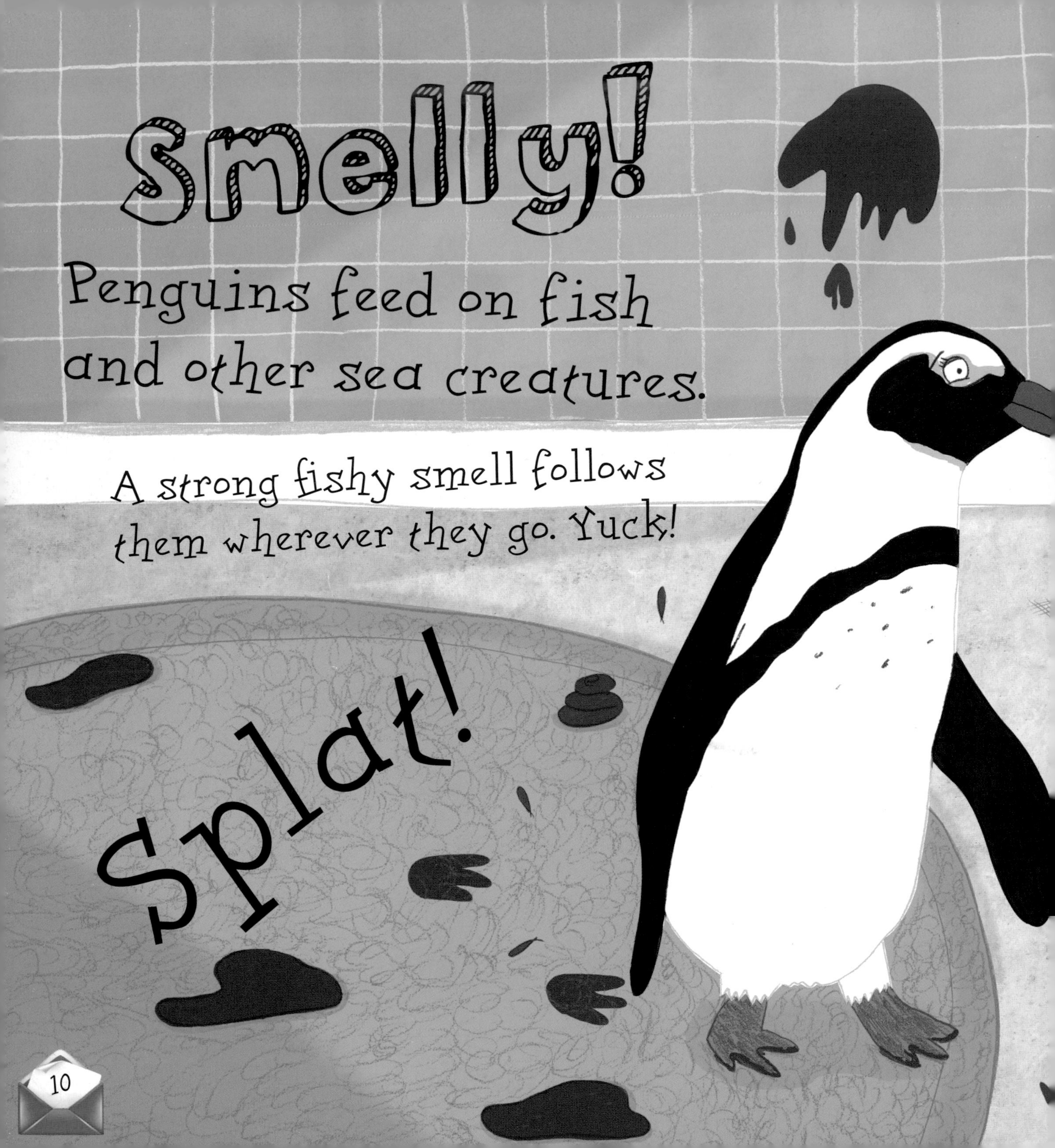

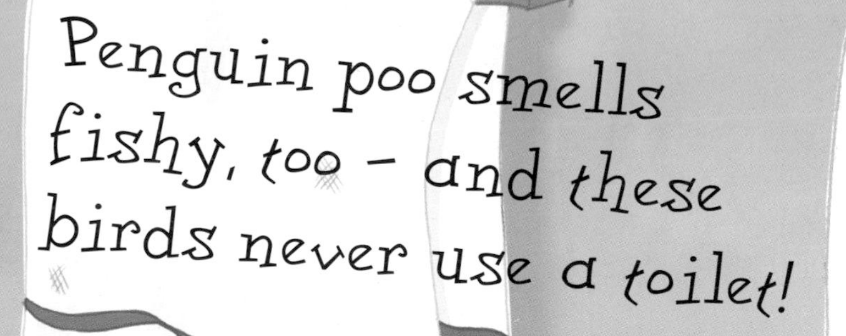

Penguin poo smells fishy, too – and these birds never use a toilet!

You will need

Perfume

Air freshener

To open all the windows

Teatime

Baby penguins are very messy eaters.

Their parents eat first, then bring the food back up to feed to the **chicks**.

Yuck!
If you join your penguin for dinner, take your own food!
You will need
A strong stomach

Slick slider

Penguins may swim very quickly, but they walk really slowly.

To get around quickly, penguins slide!

Penguins shoot across the ice by sliding on their tummies.

Wheeeee!

You will need

A sledge

A pair of skis

And a crash helmet to keep up!

Watch out!

A penguin's main weapon is its beak. It has very sharp, **jagged** edges.

Penguins use their beaks to peck and pinch.

You will need

Gloves

Boots

Protective clothes

Ouch!
Keep your legs and
feet covered up!

Time to go home

Your penguin seems happy, but your parents really aren't!

It's time to post your pet back to its real home...

A goldfish makes a great pet,
but a penguin is a WILD THING!

Cool creatures

This book featured a Humboldt Penguin.

There are 17 different **species** of penguin. They all live in the **Southern hemisphere**.

Many penguins spend three-quarters of their life in the water.

The largest penguins are Emperor penguins. They grow as tall as an eight-year-old child.

Some penguins can swim very quickly – over 30 kph! That's as fast as a cheetah can run on land.

Baby penguins are called chicks. They **hatch** from eggs and are looked after by their mum and their dad.

Different species live in different places. Some like the freezing water of the **Antarctic**. Others prefer warmer water around Australia, Africa, and South America.

Glossary

air freshener a spray that makes a room smell good

Antarctic a very cold, icy place at the far south of Earth

chicks baby birds

hatch when a baby animal breaks out of its shell

jagged rough, not even

protective keeping safe

Southern hemisphere the lower half of Earth and the countries within it

species a type of animal

Thanks for having me!

The Zoological Society of London (ZSL) is a charity that provides help for animals at home and worldwide. We also run ZSL London Zoo and ZSL Whipsnade Zoo.

By buying this book, you have helped us raise money to continue our work with animals around the world.

Find out more at zsl.org

ZSL
LONDON
ZOO

ZSL
WHIPSNADE
ZOO

Take them all home!

ISBN HB 978-1-4081-7937-6
PB 978-1-4081-7938-3

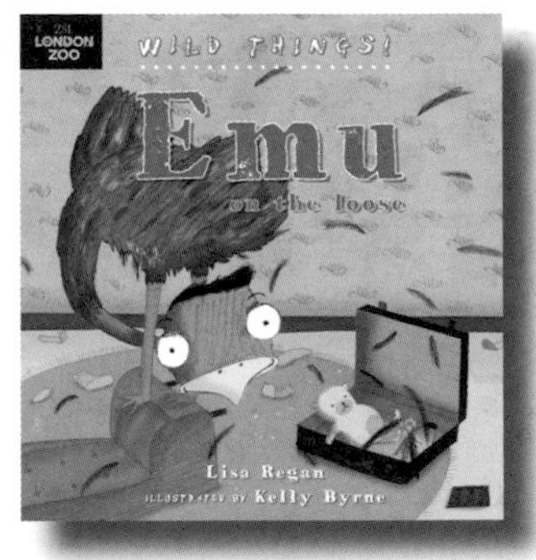

ISBN HB 978-1-4081-4247-9
PB 978-1-4081-5678-0

ISBN HB 978-1-4081-4246-2
PB 978-1-4081-5679-7

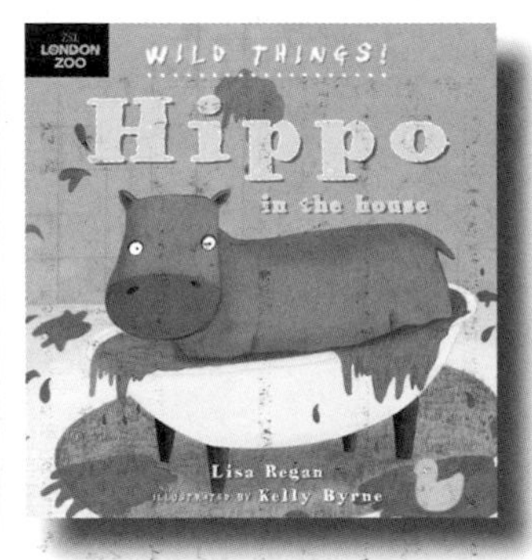

ISBN HB 978-1-4081-4245-5
PB 978-1-4081-5680-3

ISBN HB 978-1-4081-4244-8
PB 978-1-4081-5681-0

ISBN HB 978-1-4081-7939-0
PB 978-1-4081-7940-6

ISBN HB 978-1-4081-7941-3
PB 978-1-4081-7942-0

ISBN HB 978-1-4081-7935-2
PB 978-1-4081-7936-9